SPACE ELEVATOR CONSTRUCTION

Book 2 in the Space Elevator 2020 series

The concept of a Space Elevator started over a century ago but started to move from science fiction to reality in the 1990s with the promising advent of a super strong material, carbon nanotubes, that theoretically had the strength to build a ribbon into space.

NASA started investigating the concept which led to the publication of the first detailed analysis of whether it was possible, written by Dr Brad Edwards, in a NIAC paper.

Early excitement of an immediate breakthrough, maybe even constructing one by 2010, gave way to the reality of producing carbon nanotubes of the required strength, outside the laboratory, in commercial quantities. As I write, in 2020, we are still awaiting this breakthrough and the project has moved back to the late 20s, maybe the 2030s or 2040s.

But it will happen, and when it does, it will replace rockets as the easier and cheaper way to leave Earth, opening up space travel the way airplanes opened up world travel.

This series examines the technical aspects and how it will be deployed.
This second book in the Space Elevator 2020 series details the construction methodology.

CONTENTS

1. REFLECTION

In 1957...
no rocket had ever traveled to space.

Yet 12 years later in 1969...
astronauts were walking on the Moon!

Why can't you go into space?

The biggest hurdle is not space travel itself but getting off of the Earth. Or at least, once you get to space, there are new hurdles, but getting off of Earth is the first one.

You have to be age 60 or so to remember the first moon landing. The "baby-boomer" generation grew up on the Jetsons, Star Trek and Star Wars, portraying a future where space travel was routine, and huge space stations orbited the Earth.

Beam me up, Scotty? Not anytime soon.
Credit: Smithsonian Air and Space Museum

The problem is the enormous cost of leaving the Earth and getting into space. As long as rockets are the only answer, space will be for the privileged few only. At a reported price tag of $40 million just to visit the International Space Station (ISS), which at an average height above Earth of 386km is barely in space, how many people can afford it?

Each President of the USA has started their presidency by announcing some grand new scheme for American space travel, and President Trump has been no exception, calling for a Moon program, including a space station, like the ISS, orbiting the Moon. Of more concern is his call for a military force in space. But the NASA budget?

The answer is: you can have as much space travel as you like, as long as the cost doesn't exceed $19.9 Bn which is the 2019 budget, more or less the same as previous budgets.

Most of that would go toward rockets in an exploration program. If only there was an alternative and cheaper way of leaving the planet.

Well there is – or at least, an alternative is on the drawing boards, and it's not rocket science – literally!

We introduced you to it in Book One: The Space Elevator Concept.

Think of the Space Elevator as an elevator car ride, only instead of stopping at the top of a building, it continues into space, climbing up a 100,000 km ribbon!

Is it possible?

Twenty years ago we would have said no, and yet look how fast technology creeps up on us! Carl Sagan used to talk about things moving from the "impossible" to the "routine". Right now, this technology has moved over the halfway line and is being actively worked on as a realistic method of space travel. However, progress has been so fast that most people haven't even had time to absorb the idea.

We understand you feeling skeptical about the idea. Most people are at first until they hear the story and the evidence to date. Here we show you that evidence. Having said that, as authors we shake our heads at the mixed-up belief systems in the world today. So many people are gullible enough to believe in aliens, corn crop circles, horoscopes, fake news, past lives, ghosts or gremlins, shared on FaceBook, Instagram or Twitter, yet those same people will question the Moon landing and scientific facts. In this exciting world we live in, we have to continually check and question things put before us, and you are to be applauded if you approach our book with an open but questioning mind.

If only space travel was cheaper!

Sorry, but as long as we use rockets to get into space, the cost will not come down, and only a handful of lucky astronauts will ever go there. You and I are never likely to be on that journey.
The dream of the baby-boomer generation was to enjoy space travel.

For the baby-boomer generation growing up on the Jetsons, Star Trek and Star Wars, the Moon landing seemed confirmation that it was only a matter of time before everyone had access to space. It should have been the next logical step, with dreams of regular shuttle flights into orbit at a fraction of the launch cost of a Saturn V (the rocket used for the Moon landings back in the sixties).

Yet, here we are in the 21st Century, and what happened to our dreams?

For a generation that has grown up on science fiction visions of easy space access, with plenty of space stations and shuttles and rockets zooming around up there, this is a disappointment. The plain reality is that, based on present cost structures, those visions will remain just that, science fiction.

Yet that is not what we want!

If, by some magic, affordable space travel became a reality, what then? If travel to the Moon were as easy as travel by jet to a foreign country, how many of us would be booking the trip?

Get in line: we're there ahead of you!

In fact, we are ahead of you in a very significant way: we are build-ing that future!
What have we got?

2. WHY DO WE USE ROCKETS TO GET INTO SPACE?

It sounds like the sort of silly question that doesn't need asking. After all, what other way is there? It's certainly true that, right now, rockets are the only choice, but it doesn't have to be that way. As recently as 1999, we would not even have raised this question, as there was no alternative.

Your first reaction may be: "Why care about whether to use a rocket or anything else to get into space? Besides, can't I trust the experts at NASA to decide this kind of thing?"

Let's address that first question: "Why do we use rockets to get into space?"

Falcon Heavy, the world's most powerful rocket

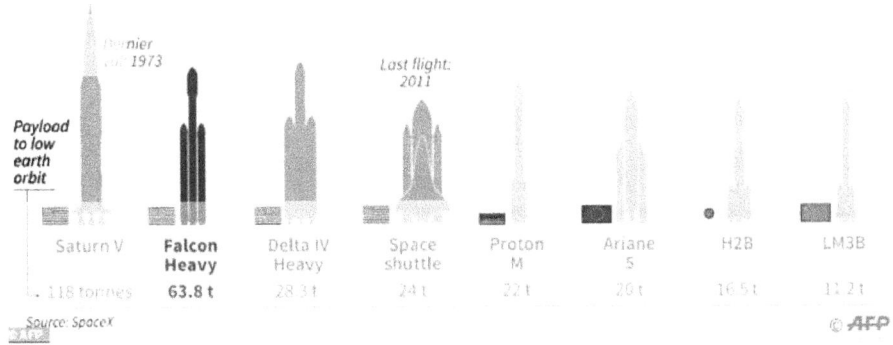

Source SpaceX, © AFP

Simply put, until now it is the only way to get into space that we have thought of. We have envisaged other ways of doing it. Science fiction writers of the past have been remarkably good at conjuring up ideas in advance of their time – even rockets were once simply a good science fiction story!

But to date, we have not managed to invent or discover the necessary technology to implement the other ideas. These ideas have included slingshots, cannons, skyhooks, space elevators, magnetic levitation, earth-based beaming propulsion systems, anti-gravity devices, wormholes, warp drives, and so on.

Rockets proved to be impossible until the 1950s. The first realistic rocket launches into space were over 60 years ago! Then the race to the Moon spawned an entire rocket-production industry. Once in existence, the momentum of production combined with the financial self-interest of people and groups involved have maintained the industry. The military applications of rockets gained appreciative customers in the biggest countries worldwide.

For space travel, rockets certainly get us there, but at an enormous cost in the order of $10,000 per kg of material launched into LEO (Low Earth Orbit), i.e. the nearest part of space! LEO is only 350 km or thereabouts above the surface of the Earth, three times the altitude of the planned sub-orbital tourist flights in SpaceShipTwo by Virgin Galactic.

The ISS (International Space Station) orbits at LEO but, so far as we are concerned, it is only just in space, not far about the atmosphere, not even out as far as the Van Allen Belts.

As a result, space is a specialist destination for major governments and satellite and communications companies. At that price it's not an everyday destination and never will be.

What about that second question: Can't we leave it all to the experts at NASA?

Such is the cost of each rocket launch that the payloads themselves are expensive, with many years of work and dollars going into extracting the most from every satellite launched. In turn, the planning and preparation required for rocket launches means that the space launch industry has the future planned for many years ahead, typically 20 to 30 years. Planning a return to the Moon has put NASA onto a timetable where events in the 2030s are already in the diary.

The current (2019) plan, favored by President Trump, as we write, is for a new space station orbiting the Moon, the Lunar Orbital Platform-Gateway, to be launched in pieces and assembly commencing in the late 2020s. Based on NASA project performance of the past, this may perhaps be optimistic. Even if the next President doesn't cancel it, it may be a 2030s project in reality. According to NASA in a February 2018 report:

'As NASA sets its sights on returning to the Moon, and preparing for Mars, the agency is developing new opportunities in lunar orbit to provide the foundation for human exploration deeper into the solar system. For months, the agency has been studying an orbital outpost concept in the vicinity of the Moon with U.S. industry and the International Space Station partners.'
'Habitation capabilities launching in 2024 will further enhance our abilities for science, exploration, and partner (commercial and international) use. The Gateway's habitation capabilities will be informed by NextSTEP partnerships, and also by studies with the International Space Station partners. With this capability, crew aboard the Gateway could live and work in deep space for up to 30 to 60 days at a time.'

'NASA's Space Launch System rocket and Orion spacecraft are the backbone of the agency's future in deep space. Momentum continues toward the first integrated launch of the system around the

Moon in fiscal year 2020 and a mission with crew by 2023.'

Image source: NASA

What has changed, in the USA, in the past decade, has been the private space industry, which is starting to supplement or even replace NASA in the launch industry.

Once in space, the rest of the space travel technology is surprisingly well advanced. What this creates is momentum and inertia: everyone gets so wrapped up in plans that typically will devote a persons' entire career to one project. As a result, less time is given to considering other ways of going about it.

In fact, the prospect of a better and cheaper technology is not always welcome: whole careers, companies and profits are predicated on the present order being maintained.

If it were possible to launch into space cheaply and regularly, the way airplanes fly every day, what would that do to our space program? It would change everything. We could be establishing a base on the Moon next year instead of in 20 years for example.

But combine the cost constraints with the long-term program imposed on NASA and we start to understand the NASA of today: it's essentially a long-term government department full of specialists devoting careers to small parts of a large picture. We don't mean to disparage NASA by that description; it's a description that would fit just about any large agency, and that's the point.

In addition, NASA has to deal with the changes in course dictated by each new President, as the strategic targets change. Give Mars priority! No, says a new President, give the Moon priority! Or maybe give unmanned missions priority! Every time the strategy changes, NASA has to ditch four or more years of detailed planning work and start again on the new priority.

What is next?

It takes time for a new idea or new technology to be accepted, but there comes a critical point where the pace of adoption becomes dramatic. In this new century there is definitely a desire for new and cheaper ways of leaving the planet and traveling into space. There are no lack of ideas, however most of them are impractical, or they don't lower the cost.

In fact, once in space we can apply many new and innovative technologies: the biggest hurdle is not space travel itself but getting off of the Earth.

The current private advances in technology by Virgin Galactic, Blue Origin and SpaceX look impressive, and we applaud them and the spirit behind them. But they are not so much "spaceships" as high flying rocket propelled planes.

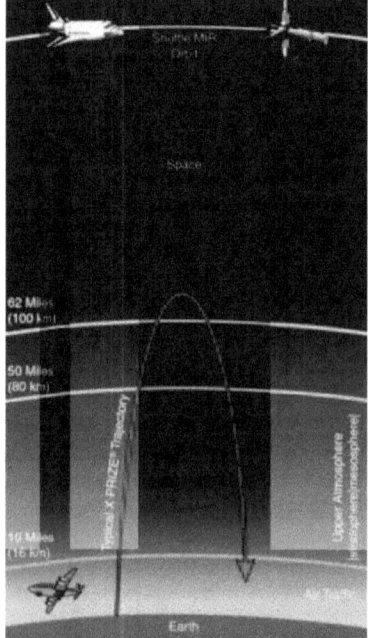

Source Universe Today

Unfortunately for these, it is not a matter of incrementally building bigger ships until they really do travel into space; they leave the atmosphere, but above 100 km the same limits apply to them as to any present-day rocket: the rules of propulsion and mass. Sub-orbital craft like SpaceShipTwo may briefly touch space for a few minutes before returning; they don't have enough power to get into orbit. To reach orbit and the ISS is ten times harder; to go to the Moon a hundred times harder. (SpaceShipTwo ascended to just over 80 km (50 mi) in December 2018, this being the lowest definition of "space" in use. A minute or two at 80 km is barely enough to call it space travel. See Book Three for a more detailed explanation of where space begins.)

We'll repeat that: you can't get into space by incrementally increasing the size of rockets. It takes a quantum leap to move from merely getting 100 km above Earth, or 350 km above Earth to get to the ISS, and leaving the planet entirely, getting out of Earths' gravity well and going to the Moon or elsewhere.

3. A SUMMARY OF THE CONCEPT

What follows is a summary presentation on the space elevator technical concepts, covered in more detail in Book One: The Space Elevator Concept.

Rockets are, at present, our only way of getting into space, due to the heavy gravity well that our planet sits in. In 1898, a Russian schoolteacher, Konstantin Tsiolkovski first speculated that rockets could be used to go into space, half a century before we had the technology to build them. Science fiction was always ahead of science fact. In some ways, it is science fiction that births the ideas and motivates people to try and make them fact. Today there are people trying to think of ways of making a warp drive, Star Trek style, though it's a long shot!

But once technology is in place, think how rapid progress can become. 1898, the idea. 1957, the first rocket was fired into space. Only 12 years later, in 1969, men were standing on the Moon. 12 years!

Same with airplanes. 1903, the Wright Brothers make the first flight. 12 years later airplanes were being used in World War One. Today we routinely fly around the world, taking it for granted, no longer marveling that we can fly to, say, Australia, in a day. Three centuries ago it was a dangerous half-year journey in a caravelle (ship).

It was the same sci-fi writer Tsiolkovski who thought of a space elevator in 1895, three years before he thought of the rocket! It lay dormant as an idea, first because you needed a rocket to get an elevator into space in the first place, second because no material was strong enough to hold its own weight. A metal tower stretching to GEO, 35,500 km away from earth, would collapse under its own weight.

In 1979, Sir Arthur C. Clarke, the famous sci-fi writer, wrote The Fountains Of Paradise, in which he imagined a space elevator tower made of some exotic metal, built in the 22nd century, based in Sri Lanka, where he lived. But it was still obviously science fiction.

Sir Arthur was once asked when he thought someone would really build a space elevator and he replied: "about 50 years after everybody stops laughing".

Well, everybody has stopped laughing, and in one of those sci-fi coincidences, we are likely to be using space elevators 50 years after he wrote his book! Sir Arthur loved that coincidence and in 2006, two years before he died, talked to Edwards and Ragan of his excitement at the work being undertaken on the space elevator concept.

What changed? It was the discovery, or invention, of carbon nanotubes (CNT) and its development in the 80s and 90s. This new material was a factor of ten stronger than any metal and led to speculation that it could be used to make a space elevator, among other things.

In the 90s Dr. Bradley Edwards was working at NASA. He is a physicist and engineer by background. NASA, from time to time, got asked to look into new ideas or concepts. Some were crazy, but they checked them out, hoping to find gold dust amongst the dross. They had a subsidiary to do this, called NIAC, the NASA Institute of Advanced Concepts. Dr. Edwards was tasked with checking out the new CNT to determine whether it could theoretically be used for a space elevator.

NASA were perhaps expecting him to reject the whole idea and say it couldn't be done. It was still regarded as something of an Indian rope trick. But Dr. Edwards did a thorough job on the analysis and he even surprised himself with the result. Theoretically, a CNT cable, extending 35,500 km to GEO, would not only be strong enough to carry its own weight, but have capacity to take payloads into space as well. His very detailed analysis, including a how-to of building a space elevator, was published in two stages as "The Space Elevator" by NIAC in 2000 and 2002.

Dr. Edwards published two more books, "The Space Elevator: A Revolutionary Earth-to-Space Transportation" with Eric Westling in 2003 and "Leaving the Planet by Space Elevator" with Philip Ragan in 2006. What followed was a wave of publicity and excitement generating many web sites, groups aiming to build it, and to regular conferences, some of which were attended by the military.

There had been a Space Elevator conference in Washington DC in 2003, and at the end, a man from the military had stood up, said that he had been impressed at the quality of work that the team had done on the space elevator, and if Dr. Edwards wanted to send up any experiments into space relating to the project, he was in a position to get spare payload allocations.

People were getting ahead of themselves, forecasting the launch of a space elevator by 2010. Unfortunately, the one holdup was the ability to produce CNT in commercial quantities. It was being produced, but only in short micrometer lengths in the lab. Optic fiber companies produce optic fiber by the hundreds of thousands of kilometers and the CNT material needs to be in production the same way, in a transfer from the lab to the production factories, in long lengths, and with absolute reliability.

NASA showed little interest in pushing the concept forward. It seems that the focus was on the rocket industry. NASA had a 30 year plan based around rockets. Plenty of firms made fortunes from rockets. No-one wanted to see a real space elevator in a hurry; one day, for sure, but after the 30 year plan.

The hardware itself is not that complicated and we have the initial designs. The plan is to get all the hardware and software ready for an earth based test run of the systems within the next decade.

So this is one idea that does have realistic prospects while also having the potential to reduce the cost of space access: the Space Elevator.

Only since about 1999 has it become apparent that it really could be built and it took the advent of a new, stronger material – carbon nanotube – to make it possible. So, it's a 21st Century idea.

Like a technologically advanced Indian rope trick, we hang a cable down from space, a ribbon 100,000 km in length. An advanced form of cable car is used to climb up and down the ribbon – no rockets!

Of course it sounds impossible, but then every new advance in technology was viewed the same way once! However, this is a serious proposition on which serious dollars are now being spent on research and development.

It is an amazing concept!

Just when you think that you understand the future and where we are going, something as incredible as the Space Elevator comes at you and changes everything.

Humanity has been here before. In fact, over a century ago the first airplanes flew, but it was hard for people to grasp the idea that, one day, airplanes would routinely fly around the world. Yet, here we are accepting air travel as an everyday event.

We confidently expect that the Space Elevator will do for space travel in the 21st century what airplanes did for the 20th century: revolutionize space travel and make it possible for anyone to venture into space.

It is now a hot topic in space travel, since it holds out the prospect of reducing the cost of travel into space by 95% or better. In short it will move space travel from the realm of government agencies, to the realm of people – to you and your children.

4. SO WHAT IS THE SPACE ELEVATOR BUILT FROM?

The Space Elevator concept is to hang a ribbon in space, reaching down to Earth, held up by angular momentum, and with cable cars climbing it by traction, much as an abseiler can climb a rope.

The ribbon rises up from the Earth's surface, and will be about 100,000 km in length, extending a quarter of the way to the Moon.

It appears to just "hang" there, but in reality it is the rotation of the Earth that produces the centrifugal forces that keep it in place.

From Earth, it will look like the ribbon goes "up" and it will be ascended by spaceships that behave like cable cars on a ski lift, crawling up the ribbon. Once a cable car is a significant distance from Earth however the lack of gravity makes it easier to view it as a "road" or "train line" into space along which the cars are traveling. The car simile is appropriate in another way too: they will be traveling into space at a speed of about 200 km/hour, the speed of a fast car. It may not be as fast as a rocket, but it will likely be only 2% of the cost!

Once it is deployed cable cars climb it using traction, to GEO (Geostationary Earth Orbit) and beyond.

The ribbon needs to be deployed into space in the first place, and for that we need to launch drums of ribbon using heavy lift rockets. The drums and associated hardware are propelled 35,500 km out into space to GEO (geostationary orbit) and from there the cable is wound out until it forms a single 100,000 km length. 35,500 km hangs down towards Earth, while 65,000 km stretches further out into space. It is not "standing" on Earth. It is not built "up" from Earth. Rather it hangs "down" from space, kept in place by the centripetal force exerted by the rotation around the Earth, the same way a rope stays in a straight line when you swing it around you.

Asked when a Space Elevator would really get built, Sir Arthur C. Clarke, the eminent science fiction writer, famously replied "about 50 years after everybody stops laughing". Now they have stopped laughing. In the time it has taken for the smartphone to be created and get accepted, engineers have come to accept that a space elevator can be constructed.

The space elevator ribbon will stretch about a quarter of the way to the Moon. Can't we just extend it to the Moon? No, for reasons we'll explain later. As you see every time you look up at the Moon in the sky, it doesn't stay in the same place, so it moves relative to a space elevator ribbon and can't be attached to it.

It needs to be built from something incredibly strong, which turns out to be carbon nanotubes (CNT).

5. CARBON NANOTUBES (CNT)

As we explained in our first book in this series, in the 20th Century the Space Elevator was science fiction. The problem was simple: no known material was strong enough to build it with.

Those calculations of material strength were done a half-century ago, even before the first rockets entered space, and it was a depressing outlook. It appeared as though mankind was forever doomed to stay on this planet. Thankfully, we have broken that constraint, even though it has taken risky and expensive rockets to do it.

Until recently no material existed that was strong enough to build the Space Elevator. In the 1990s, the discovery of strong and light carbon nanotubes, with the potential to create material 100 to 180 times stronger than steel by weight, changed everything.

But if it is that good, why hasn't someone already built the Elevator, you ask? Right now, carbon nanotube based materials have been made that are about 100 times steel in strength, sure enough, so we've proved the material can be created. But it's still in the laboratory and production methods have not evolved to the point where we can produce it in commercial lengths, as yet (2019). Once we can, the game changes.

6. UNDERSTANDING GRAPHENE

CNT is a form of graphene. Here we expand on the graphene family, as the various iterations are useful in diverse situations.

'The superstrong, superthin, and superversatile material that will revolutionize the world' is how Les Johnson and Joseph E Meany describe it in their book Graphene (Published 2018 by Prometheus Books), to which you are referred for more detail on Graphene.

Graphene is made from Carbon, the fourth most common element, so abundant as to be taken for granted. It's the same with sand on the beach. Humans sat on the sand for thousands of years before discovering how it could be transformed into glass, concrete or silicone. For just as long, there was no understanding of the link between the graphite used to write with, in old fashioned pencils, and coal, charcoal, soot and highly valued diamonds, yet they are different forms of the same material. And, by the way, our life forms on Earth, including us, are carbon based.

Graphite can be mined in abundance; it's the precursor of graphene, but it's only in the past half-century we discovered the many forms of graphene. The structure of graphene is a one-off in the world of chemistry. Johnson and Meany describe it as:

'Graphene is a perfect anomaly in the world of chemistry - a flat, two-dimensional molecule, with a single sheet of graphene measuring only one atom thick. You might immediately question the structural integrity of graphene due to its delightfully simplistic construction, but the weaving of the carbon hexagons throughout the structure makes the atomically thin material unexpectedly strong.'

The family of materials includes graphene, graphane, and fullerenes. In their raw form they are not much to look at, being mostly grey powder. But the differences come from the way carbon molecules are joined together.

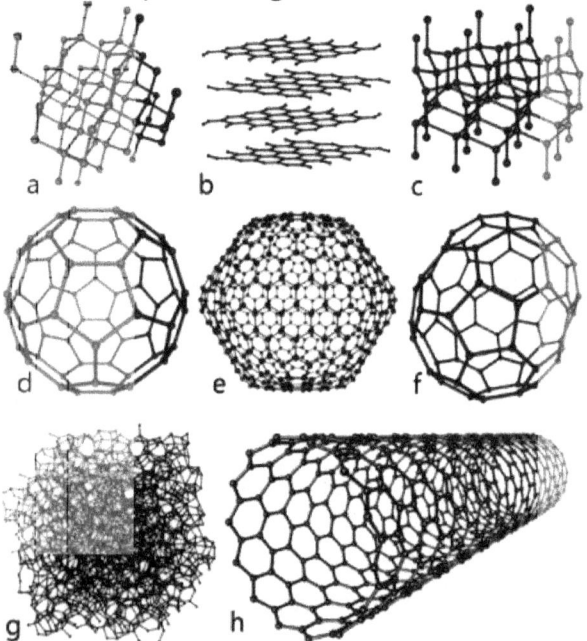

Graphic source Wikipedia

Some allotropes of carbon are: a) diamond; b) graphite; c) lonsdaleite; d–f) fullerenes (C60, C540, C70); g) amorphous carbon; h) carbon nanotube (Source Wikipedia)

Individual carbon atoms are simple in structure, with 6 protons, 6 neutrons (some forms have up to 8 neutrons) and 6 electrons. It is nonmetallic and tetravalent, making four electrons available to form covalent chemical bonds, which is when it gets interesting. The difference between them lies in the way the atoms connect to one another. These differing forms are called allotropes of carbon.

It is because carbon has 6 electrons, not 10, that it can form allotropes. The 6 electrons are located in shells, with 2 in the first shell closest to the nucleus, and 4 in the second shell. But the outer shell obeys the octet rule. There are 8 spaces for electrons in the outer shell but only 4 are occupied, so it can share the empty slots by forming bonds with up to 4 other atoms.

'The atoms of carbon can bond together in different ways, termed allotropes of carbon. The best known are graphite, diamond, and amorphous carbon. The physical properties of carbon vary widely with the allotropic form. For example, graphite is opaque and black while diamond is highly transparent. Graphite is soft enough to form a streak on paper, while diamond is the hardest naturally occurring material known. Graphite is a good electrical conductor while diamond has a low electrical conductivity. Under normal conditions, diamond, carbon nanotubes, and graphene have the highest thermal conductivities of all known materials. All carbon allotropes are solids under normal conditions, with graphite being the most thermodynamically stable form at standard temperature and pressure. They are chemically resistant and require high temperature to react even with oxygen.' - Wikipedia

7. CARBON ALLOTROPES

What we are interested in is the crystalline allotropic forms of carbon, of which the first to be recognized was graphite, mined from coal seams. So-called "lead pencils" do not contain lead, but rather a mixture of graphite and clay. Graphite is electrically conductive and can be used as a lubricant.

Graphite has a layered, planar structure and each of the layers are called graphene.The carbon atoms in each layer are arranged in a honeycomb lattice, bonded covalently, with only three of the four potential bonding sites used. The fourth electron is free to migrate in the plane, making graphite electrically conductive. Interestingly, it does not conduct in a direction at right angles to the plane.

The next interesting structure to be discovered was Buckminsterfullerene, or, as they are sometimes called, buckyballs, because of the polyhedral carbon structure composed of around 60-80 carbon atoms in pentagon and hexagon configuration. Discovered in 1984/5, they are named after Buckminster Fuller and have a structural resemblance to geodesic domes or spherical soccer balls. As a kind of empty ball, Fullerenes are used for drug delivery systems in the body, in lubricants and as catalysts. Connected together in tubes, they were the first items called nanotubes. They are electrically conductive.

Graphene is electrically conductive, but it can be linked with hydrogen in a process called reduction. The two dimensional crystal is reduced to graphane, with each carbon atom linked to a hydrogen atom. Graphane is non-conductive. The C-H bond is not strong and one use suggested has been to use graphane as a way of storing hydrogen in fuel cells, where the hydrogen is released, turning the graphane back into graphene. This may have a use in space as a secure method of storing hydrogen in fuel cells.

Carbon nanotubes (CNTs) are allotropes of carbon with a cylindrical nanostructure, having unusual properties, which are valuable for nanotechnology, electronics, optics and other fields of materials science and technology. With exceptional strength and stiffness, nanotubes have been constructed with length-to-diameter ratio of up to 132,000,000:1, significantly larger than for any other material.

CNT's are sheets of graphene rolled up on themselves, like a tube, made of the same hexagonal structures. Each end is capped with hemispherical structures, in effect, half a buckyball at each junction, forming a ridge along the rim. CNT's are about a million times longer than they are wide and in practical terms are considered to be one-dimensional materials.

Above, we noted that graphite is made up of graphene layers. The breakthrough was achieved in realizing the value of separating these layers, to be able to hold just one of those layers. One layer is monolayer graphite, now called graphene.

Here's an experiment you can do at home, under adult supervision. Light a candle, then hold some glass above the flame, taking care not to burn yourself. The flame will deposit a thin black layer of soot on the glass. You have just created graphene! If you didn't know better, you could be forgiven for thinking this black layer had no particular use, but you are looking at the material from which the space elevator ribbon will be made (with a few production tricks along the way).

So graphene was known about for much of the last century, but uses for graphene were not obvious, until the 1990s. The research done by Edwards at NIAC in the late 90s marked a turning point in graphene research and the number of research papers on graphene has grown exponentially since then. But there are still challenges and opportunities ahead. We are interested, less in research, than in getting the material we want into production, with a lack of bulk production as yet.

8. THE STATUS OF NANOSCIENCE

In a paper published in the journal Carbon, volume 98, in 2016, "Carbon science in 2016: Status, challenges and perspectives", the authors wrote, inter alia:

'Covalently bonded nanocarbon-based materials'

'During the last 25 years, new allotropic forms of pure carbon, mostly sp2 bonded with nanoscale dimensions (e.g. fullerenes, nanotubes, graphene), exhibiting novel physico-chemical properties, appeared thus triggering intense research in the area of carbon nanoscience. These unique structures have been at the forefront of nanotechnology and have had their impact as the most important building blocks in future nanotechnology applications. However, Carbon researchers should now start developing the ability to scalably build well-ordered three dimensional (3D) bulk materials using these nanoscale building blocks of carbon, using systematic and efficient approaches. The implications of this are tremendous: 1) no such defined material at bulk state has been reported even though many heterostructures have been observed at the micro scale; it may be equally important for the truly scalable/bulk applications even at the bulk state of these structures, particularly for CNT and graphene, since stacking/aggregation of CNT and graphene has been hampering their bulk applications; 3) some new and unprecedented properties or phenomenon, otherwise not achievable for carbon solids today, may be produced or observed with these heterostructure based materials. The challenge would be to create such engineered 3D nanostructured carbon materials with controlled (and repeatable) structure units at bulk state.'

("Carbon science in 2016: Status, challenges and perspectives" by Jin Zhang, Mauricio Terrones, Chong Rae Park, Rahul Mukherjee, Marc Monthioux, Nikhil Koratkar, Yern Seung Kim, Robert Hurt, Elzbieta Frackowiak, Toshiaki Enoki, Yuan Chen, Yongsheng Chen, Alberto Bianco.)

We are amused by their reference in the paper that, in 2016, "nanoscience is maturing as a field and the graphene discovery is ten years old." Such is the pace of R&D these days, that a ten year old field can be described as a maturing field!

Carbon nanotube is just that, a tube, grown in a continuous thread by adding carbon atoms at the end of the mat. So how do we create a ribbon out of tubes? The tube can be considered as a string, wire or strand of wool. Tubes can then be knitted together to form a mat or ribbon.

Quoting from the paper again:

'Most of the materials that have interested carbon science researchers to date are built with the benzene ring, which consists of an sp2-bonded hexagon, as the primary building block. Fusing an infinite number of benzene rings in two-dimensional manner gives graphene. Stacking graphene sheets gives graphite. A carbon nanotube is made by rolling up a semi-infinite graphene sheet to a hollow cylinder, while adding pentagon rings into a graphene nanofragment in building a sphere brings about a fullerene. By cutting a graphene sheet into nanofragments, nanographenes such as graphene nanoribbons are created.'

'High-purity single-walled carbon nanotubes: growth, sorting, and applications'

'The single walled carbon nanotube (SWCNT) is a carbon allotrope that has a unique 1D tubule structure formed by wrapping a single graphene sheet. The twist angle of the graphene sheet and the diameter of tube defines the SWCNT's chirality, which controls its electronic, optical, thermal, mechanical and magnetic properties. Since the first report of SWCNT synthesis in 1993, SWCNTs have always been produced as a mixture of different chiralities, and thus obtaining high-purity SWCNTs has been a key hurdle for realizing their wide potential applications. In the last three years, research efforts from different fronts have reached a point where high-purity SWCNTs are on the horizon.'

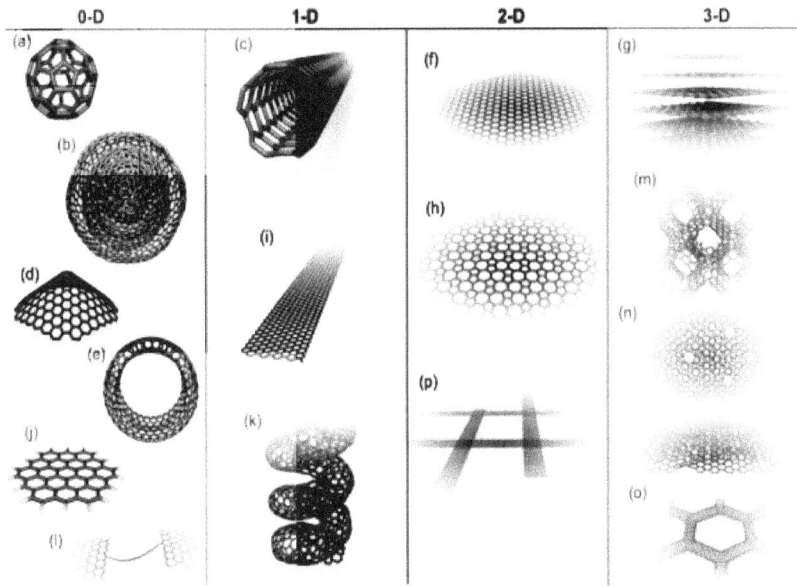

Image from Terrones et al, sourced from the same paper.

Molecular models showing different types of carbon nanomaterials categorized by their dimensionality. It includes fullerenes, nanotubes, nanoribbons, graphene, nanotube networks, etc. that could be used as building blocks (units) to synthesize real 3D carbon materials with controlled structures and properties. For a space elevator ribbon, we are synthesizing in the form (C) shown in the image.

However, the research work on producing nano carbon composites continues to challenge, as the paper goes on to state:

'Nanocarbon materials represented by carbon nanotubes and graphene-based materials are being widely explored in fabricating nanocomposites with an expectation to impart excellent mechanical/electrical/thermal properties to the composites. Indeed, numerous studies have emphasized the excellence of nanocarbon materials with respect to a variety of properties …'

'However, in reality, the performance of nanocomposites has not yet reached the level of expectation. For example, the mechanical performance of nanocarbon-reinforced nanocomposites is less than one third or half of that of nanocarbon materials themselves …'

'It is nonetheless rather frustrating to find that the performance of the nanocomposites is still below our satisfaction despite such numerous efforts to improve the dispersibility and the interfacial interactions of nanocarbons. Further, many of the reports of improved performance are not reproducible even when strictly following the published recipe and procedures. This situation may be one of the reasons hindering the widespread commercialization of nanocomposites …'

'The grand challenge in the carbon nanocomposites field is to fabricate materials that achieve ultimate performance by fully utilizing the intrinsic properties of nanocarbon fillers.'

Although good work is being undertaken in some labs, this still sums up where we are today. We have yet to demonstrate a reliable method of bulk production of graphene, of the desired strength. But, as the paper reminds us, the field is only just over a decade old, and we look to the longed-for breakthroughs in the next decade.

9. GRAPHENE AND ELECTRICAL CHARGE

Graphene conducts electricity, having low resistance to current flow, which is useful for some purposes but not for the space elevator ribbon. Ideally, the ribbon should be non-conductive. This causes two types of problems to manage. At sea level, in the atmosphere, it could theoretically collect a static charge, or a surge from a lightning strike, which could be transmitted along the ribbon into space, and particularly in the case of lightning, could melt the ribbon with adverse consequences. In space, it could collect a charge from the planetary magnetic field (as in a standard generator which converts magnetism to electricity), or pick up a charge from the solar wind.

On Earth, we mostly use copper or aluminium to transmit electricity, as the cost effective choices. Copper rates as the most conductive, with aluminium less conductive. Graphene is 1.4 times as conductive as copper by volume, but when density is factored in, it is 5.8 times more conductive (if a given sample of copper can carry 1 kA with a 1 V drop in voltage, the same weight, and length, of graphene, could carry 5.8 kA with with a 1 V drop in voltage - in theory, assuming it is a sheet of perfect graphene). Of course, fabrication methods can degrade this. For electrical power transmission, which depends a lot on weight, aluminium is twice as conductive as copper by weight and so is used for long distance power transmission. (Stack Exchange)

However, just as for copper and aluminium, power transmission degrades with distance. In simple terms, a higher voltage can be transmitted over a longer distance along a wire, which is why the cables from a power generating station are massive and carry voltages of 110,000v or more, unlike the 110v to 240v in your home.

10. THE SPACE ELEVATOR AND ELECTRICAL CHARGE

The ribbon is modeled at a length of 100,000 km. Most of the static electricity referred to above comes at relatively high voltages, in the range of 20,000v to 40,000v, but has low amperage … in simple terms, it comes in low quantities.

The ribbon passes through the Van Allen Belts, which is a zone of high radiation and electrical charge. Could the ribbon short out the Belts and create a current between Earth and the Belts? We think not. The electrical charge of the Belts is large based on the entire volume of the belts, but space is big, and the charge around the relatively small ribbon would be measurable but not likely to have a detrimental impact. The ribbon is a few meters in cross-section. Beyond the ribbon itself, it is not measurably affecting any electron or ion whereas an electron has to hit the ribbon for any grounding to occur. In numbers: the Van Allen Belts have an volume in the order of 10^23 cubic meters. The ribbon cross section is about 10^5 cubic meters which is some 13 orders of magnitude difference.

A single ribbon into space does not create a circuit for electricity to flow, but some from of two-ribbon system would have electric potential. We won't be sure how much electricity is involved until we can manufacture ribbon and test it, on the ground and in space.

At first, utilizing the conductivity might appear to be a way of sending electric power up the ribbon. It would be neat if we could power the cable cars via this system. But there is a relationship between the cross-section of a ribbon and the distance involved. While graphene conducts electricity, the material has resistance causing power losses, just like copper cables, imposing a limit on distance, how far useful electricity can be transmitted. Subject to testing this in practice, we don't expect the ribbon itself to be a useful transmitter of power.

It's no good running a set of copper wires up the ribbon, the sort of thing you use at home, either. The further the distance electricity has to be transmitted, the higher the voltage needed to push it along the wires and the larger the wires have to be. Your house might use 110 volts (USA) or 220/240 volts (rest of world), but the power company uses large overhead lines at anything from 60,000 volts to 765,000 volts to send power from the power station to your home, using transformers to step the voltage down to your home voltage.

Within the atmosphere, we need to test and manage any buildup of electrostatic charges. These attach to any body moving through the atmosphere, even airplanes. When flying, you may have noticed wires sticking out at the end of the wings. These serve the purpose of bleeding the electrostatic charge as the airplane travels through air, and the cable cars will likely need similar.

The good news is, any electric charge generated at the Earth end will dissipate along the ribbon due to the same power losses. Even so, care is needed with static electricity in the presence of oxygen; we don't want fuel ignited. Also, a lightning strike on a ribbon, besides the risk of melting the ribbon, might induce a high power surge along the ribbon, which is the main reason we want the ribbon positioned at places on Earth where lightning is almost non-existent.

It is possible for electrostatic charges to build up as the cable car travels through the Van Allen Belts, too. This will require modeling and testing in-situ, once CNT ribbon is available for test deployments.

In terms of considering the structural integrity of the ribbon, the significant risk comes from a lightning strike, where the total current is large enough to be destructive.

Ways of managing the lightning risk have been considered and will require testing when the CNT material is available. These include passive and active systems:

* Selecting Earth locations where the risk of lightning is minimal to non-existent.
* Erecting high lightning rod conductors near the ribbon to deflect lightning (passive)
* Shielding the CNT graphene with graphane or other insulator for the lower 10 km of the ribbon (passive)
* Disconnecting the ribbon from the Earth Space Port when a storm threatens, winding it up to clear the atmosphere till the storm passes (active)

A conveyor-belt style ribbon (covered in the Ribbon Design section) poses additional problems with lightning as some of these approaches can't be used.

11. USING GRAPHENE ON EARTH

While we have focused on the use of graphene-based CNT for the space elevator, the material will first get used for applications here on Earth and in fact is already in use. This has a couple of advantages, aiding the expansion of research into graphene, and in proving up the strength of the material before we consider launching it into space.

Although the graphene field is fairly new, having been around for a couple of decades, the application of new technology happens quickly in this century. Last century even, it took only a decade for nuclear scientists to turn paper calculations into an atomic bomb. It took 12 years to develop rockets to the point where we landed on the Moon.

Right now, it seems like new research on graphene is being published daily, such is the momentum. The potential applications of graphene include:

* Nanotechnology in medicine, medical telemetry.
* Consumer electronics (touch screens, integrated circuits.)

As a composite material it has found applications in:

* Construction
* Automotive bodies
* Energy (distribution, cables)
* Airplane manufacture

* Military vehicles

A few applications will have relevance to the space elevator, pro-viding ways of testing the material, its strength, capacity and management needs.

One application will be in existing cable car technology, as used in ski lifts, replacing the metal cables with CNT cables. The po-tential is to install ski lifts to heights not feasible at present. Imagine a ski lift to the top of Mount Everest, Mount Hood or similar.

Suspension bridges are currently built with steel, which is strong but heavy. An early use of the long-chain CNT material will be to erect longer and lighter suspension bridges, an application for which CNT will be ideal. We can also test non-conductive coat-ings in-situ to manage lightning risk.

Suspension bridges will be an exciting application. Imagine a road and rail suspension bridge crossing the English Channel from Dover to Calais, or supplementary bridges adjacent to the Golden Gate bridge, or joining Brooklyn to Manhattan. The potential length of suspension bridges will be limited, not by the weight of steel, but by the curvature of the Earth, allowing for bridges of up to 40 km in length.

When suspension bridges have been built with CNT and proven to be reliable, it will provide the impetus to commence the first space elevator.

Of course, a more mundane application will be for existing lift shafts in buildings. The lifts or elevators in buildings can be sus-pended by CNT instead of steel. This gives the potential for taller buildings serviced by lifts or elevators beyond currently feasible limits.

12. STATUS OF CNTS FOR THE SPACE ELEVATOR

So where are we at with CNT production? The Space Elevator is unique in that it requires material much stronger than any we have available. We need material that is 100 to 180 times stronger than steel, 5 times stronger than our current best carbon fibers. Fortunately for us the material we need exists, though not yet fully developed for commercial use.

Carbon nanotubes (CNT's) are a very specific form of tubular carbon molecule. It is the strongest material that has been made or will ever be made - assuming our physics is correct. These carbon nanotubes have been measured at strengths 400 times stronger than steel for the same mass - at first glance, well beyond what we need. Just amazing!

The nanotubes in current production are very short, not nearly long enough to reach 100,000 km. Gluing them together to make a long ribbon is possible, of course, but then the strength of the cable would be limited to the strength of the glue - not a solution. What we need to be able to do is to tightly spin the nanotubes into fibers and then into entire 100,000 km lengths.

This is the most significant hurdle holding back the Elevator right now and it is generally considered that it will take another decade of R&D before we achieve the required strength.

However, research on nanotubes is being carried out by numbers of competing groups and it is only a matter of time before we will be able to grow nanotubes any length we want and combine them into long threads.

The most advanced work publicly revealed is one published in May 2018 by Rufan Zhang of Stanford University.

The work has been carried out by Zhang, Fei Wei, Xuan Ye and colleagues at Tsinghua University in Beijing, China.

They seem to be leading CNT development at the moment, growing long (up to 55cm) and very strong CNTs (80GPa). They write:

'Carbon nanotubes (CNTs) are one of the strongest known materials. When assembled into fibers, however, their strength becomes impaired by defects, impurities, random orientations and discontinuous lengths. Fabricating CNT fibers with strength reaching that of a single CNT has been an enduring challenge. Here, we demonstrate the fabrication of CNT bundles (CNTBs) that are centimeters long with tensile strength over 80 GPa using ultra-long defect-free CNTs. The tensile strength of CNTBs is controlled by the Daniels effect owing to the non-uniformity of the initial strains in the components. We propose a synchronous tightening and relaxing strategy to release these non-uniform initial strains. The fabricated CNTBs, consisting of a large number of components with parallel alignment, defect-free structures, continuous lengths and uniform initial strains, exhibit a tensile strength of 80 GPa (corresponding to an engineering tensile strength of 43 GPa), which is far higher than that of any other strong fibre.'

("Carbon nanotube bundles with tensile strength over 80 GPa." Bai, Yunxiang & Zhang, Rufan & Ye, Xuan & Zhu, Zhenxing & Xie, Huanhuan & Shen, Boyuan & Cai, Dali & Liu, Bofei & Zhang, Chenxi & Jia, Zhao & Zhang, Shenli & Li, Xide & Wei, Fei. 2018.)

The 'Daniels effect' refers to a paper titled 'Strain Rate Characterization of Unidirectional Graphite/Epoxy Composite' by IM Daniel, WG Hamilton and RH LaBedz (American Society for Testing and Materials, 1982, pp 393-413), which reports, inter alia:

'In testing composites at high strain rates wave propagation effects become dominant and cause difficulties in the interpretation of results. To minimize these effects a new method for testing composites ... utilizes a thin ring specimen loaded with a dynamic internal pressure pulse. The transit time of the stress wave across the thickness of the specimen is very short compared with the duration of the test. This paper describes the application of this method for testing and characterization of unidirectional graphite/epoxy over a wide range of strain rates.'

Brain Wang in nextbigfuture reports on the paper as follows:

'Breakthrough carbon nanotube bundles are over 20 times stronger than Kevlar'

'Finally ultra-long (several centimeter) carbon nanotube fibers have been made into stronger bundles. The tensile strength of CNTBs (Carbon nanotube bundles) is at least 9–45 times that of other materials. If a more rigorous engineering definition is used, the tensile strength of macroscale CNTBs is still 5–24 times that of any other types of engineering fiber, indicating the extraordinary advantages of ultra-long carbon nanotubes in fabricating superstrong fibers.'

'Superstrong fibers are in great demand in many high-end fields such as sports equipment, ballistic armour [sic], aeronautics, astronautics and even space elevators. '

'In 2005, the US National Aeronautics and Space Administration (NASA) launched a 'Strong Tether Challenge', aiming to find a tether with a specific strength up to 7.5GPa cm3 per gram for the dream of making space elevators. Unfortunately, there is still no winner for this challenge.'

'The specific strength of existing fibers such as steel wire ropes (about 0.05–0.33 GPa cm3 per gram), carbon fibers (about 0.5–3.5GPa cm3 per gram) and polymer fibers (about 0.28–4.14GPa cm3 per gram) is far lower than 7.5GPa cm3 per gram).'

'Carbon nanotubes, with inherent tensile strength higher than 100GPa and Young's modulus over 1TPa, are considered one of the strongest known materials.'

'Generally, there are three types of CNT:

* agglomerated CNTs
* vertically aligned CNT (VACNT) arrays
* ultra-long horizontally aligned CNT (HACNT) arrays ('ultra-long CNTs' for short).

'Almost all the reported CNT fibers are fabricated using agglomerated CNTs or VACNT arrays with lengths less than a few hundred micrometers and with plenty of structural defects and impurities, giving those CNT fibers a tensile strength ranging from about 0.5 to 8.8GPa which is much lower than that of single CNTs.

'Ultra-long CNTs should have great advantages in fabricating fibers because of their macro scale lengths (ranging from centimeters to decimeters), neat surface, perfect structures and super-parallel alignments. But because the production of ultra-long CNTs is extremely low, there have been no reports of fibers fabricated using ultra-long CNTs, so the question of whether ultra-long CNTBs possess equivalent strength to single CNTs has remained open.'

'Researchers have fabricated CNTBs that are several centimeters long, using ultra-long CNTs with defined number and parallel alignment, to quantitatively investigate the relationship between the tensile strength of ultra-long-CNT-based fibers and their components.'

'Generally, the ultra-long CNTs are synthesized through a gas-flow directed chemical vapor deposition (CVD) method with parallel orientations and large intertube distance on flat substrates. The resulting CNTs usually have one to three walls with perfect structures.'

A news report in the South China Morning Post followed it up with:

'A research team from Tsinghua University in Beijing has developed a fiber they say is so strong it could even be used to build an elevator to space. They say just 1 cubic centimeter of the fiber – made from carbon nanotube – would not break under the weight of 160 elephants, or more than 800 tonnes. And that tiny piece of cable would weigh just 1.6 grams.

'This is a breakthrough,' said Wang Changqing, a scientist at a key space elevator research centre at Northwestern Polytechnical University in Xian who was not involved in the Tsinghua study.

The Chinese team has developed a new 'ultralong' fibre from carbon nanotube that they say is stronger than anything seen before, patenting the technology and publishing part of their research in the journal Nature Nanotechnology earlier this year.

'It is evident that the tensile strength of carbon nanotube bundles is at least 9 to 45 times that of other materials,' the team said in the paper.'

13. CONSTRUCTING A SPACE ELEVATOR SYSTEM

We have the vision. Now we have to turn it into reality. Our immediate goal is to get the first Space Elevator constructed and operational sometime in the 2030's, with further Elevators under construction thereafter.

The process and the steps needed have already been defined, and will be explained in more detail, following. Here they are here explained briefly.

First, we need to solve that initial problem of the required material. Steel, silk, Kevlar, carbon? None of them come close. Only carbon nanotubes are strong enough and light enough. They are available commercially – you can go on the Internet and buy samples yourself. The problem is combining the nanotubes into reliable large-scale materials that can be used to make fibers and eventually a Space Elevator cable.

Having the material ready is just the start. The various components will need designing and testing. Designs are being developed for the Elevator Space Port, for the cable cars that will ride the cable, and for anchor weights and space stations at the far end of it.

Spaceships will still need rocketry to leave the far end of the cable and travel to the Moon and other destinations, and these will need to be designed and tested. General design work will need completing by 2025.

14. THE ELEVATOR SPACE PORT

There needs to be an Earth spaceport, most likely a terminal floating on the ocean, to which the Space Elevator cable will be attached.

Where will this Earth base or terminal be situated? A number of locations have been canvassed, and the restrictions on choices are not as tight as was thought a decade ago. These are:

* On, or south of, the equator south of Hawaii
* South of the equator, in the Pacific Ocean at some point between Fiji and Karajini.
* The southern Pacific off of the coast of South America.
* The mid-Atlantic off the coast of South America, around French Guiana.
* The Indian Ocean off the coast of Western Australia.

Currently, the favored locations for the Elevator Space Port are south of the equator in the Pacific Ocean. (See Book Four in this series for more details.)

All of these sites are on or south of the Equator. For reasons explained later, there are no suitable sites north of the Equator. A key factor is to locate a place which is calm, weather-wise, preferably with no storms, no hurricanes, no thunder or lightning. There are around eight locations on Earth which meet these requirements, but other issues also come into play.

The floating terminal (think JFK or Heathrow in size) will need to be massive enough to resist the outward (i.e. "upward") pull of the cable and be large enough to potentially support a floating airport, with housing, a construction facility and a boat harbor.

15. CABLE CARS

The ribbon will be fixed in position, with "cable cars" or crawlers climbing up and down it.

The cable cars will have more in common with today's jumbo airplanes than with rocket modules. As they ride up the ribbon, the apparent weight diminishes and riders transit to experiencing weightlessness. The cars will have grips and wheels that clamp on the ribbon to climb it, using principles established from ski lift technology by several commercial applications. The cable cars will need to be light, but unlike cabin space in rockets, they can have a large volume, allowing people to float around in the weightlessness of space.

16. DEPLOYING THE RIBBON

An important part of the process will be to test the ribbon and cable cars in space, by laying out sections of ribbon in a tether arrangement. The ISS could be used as the "hook" from which to hang up to 250 km of cable. It is not possible to construct a ribbon from the ISS to the surface of the Earth, since the Earth end would be whizzing by at 22,000 km per hour and burn up. However, with a 250 km length suspended in space from the ISS, we can test the ribbon, its strength, and various competing climber technologies.

The next significant test would involve sending a payload of ribbon 35,500 km out into space, to geosynchronous orbit ("GEO"). Here we can practice reeling out ribbon by the thousands of kilometers while learning how not to get it tangled!

The method of deployment calls for using rockets to launch spindles, on which the entire 100,000 km of ribbon is rolled up. Launched first into Low Earth Orbit ("LEO") by rocket, the spindles and associated control equipment will then be powered by rocket up to GEO, 35,500 km away.

Once in stable orbit, the spindles are lined up so that the ribbon can be spun out, managed by computer, with the length balanced in terms of gravity and rotational forces. On completion there will be a ribbon stretching from the ground, out past GEO and continuing for 100,000 km into space. At this length, the angular momentum of the ribbon is enough for it to be suspended in place. From the Earth it will look as though it is hanging in mid-air, held up by nothing.

It is not built up from Earth. Rather, it is hanging down from space. As it is dropped into the Earth's atmosphere, we will have to "catch" the free end and drag it over to where the fledgling Elevator Space Port is floating. Once attached to the terminal, we can start to use it.

Adding strands to the ribbon

This deployment lifts only the first small ribbon of carbon nano-tube, perhaps a mere 20 cm wide; a thin, semi-transparent sheet of parallel fibers smaller than a human hair.

Now, in order to be able to lift useful masses of material, it is necessary to increase the size of the ribbon by stitching additional ribbons to it, only now there is no need to use rockets anymore!

Specially designed crawlers will ascend the initial ribbon, laying out more small ribbons behind them. Every time another strand is carried into place, the carrying capacity of the total ribbon increases, and so the speed of deployment increases. Eventually, some 200 or so ribbons will have been added, turning it into a ribbon several meters wide, strong enough to carry the full size cable cars.

So far, all of those crawlers will have been on a one-way journey to the top of the ribbon. For now, they are left there, since they form a useful mass, without which the ribbon would have to be even longer. (The length of 100,000 km has been chosen based on an engineering trade-off of ribbon length and mass compared to the counterweight needed at the upper end and how far payloads can be thrown.)

Now, for the first time in history, Earth is linked to space by a ribbon!

17. SYSTEM DEPLOYMENT ISSUES

While the Space Elevator is a complex system of many interacting and balancing parts, the deployment concept itself is fairly simple. A ribbon is placed in space and one end is lowered down and attached to Earth. Mechanical climbers can then ascend this physical bridge between Earth and space.

Of course, there are basic questions that immediately arise:

* How does the ribbon stay in place?
* How long is the ribbon?
* How do the climbers work?
* Why would we want to build this?
* What if an airplane hits it?

These are just the first of many questions. Here we will start with the technical ones and try to give you a good understanding of how the Space Elevator works. From here you will have a good foundation to see why the future rides on this system.

The initial issue is the deployment of the Elevator ribbon and the satellite (Geo Station) controlling this.

In one science fiction novel written on the Space Elevator an asteroid is captured and brought into geosynchronous orbit. This asteroid is mined to produce a cable ten meters in diameter that is extended down to Earth. These are difficult tasks to say the least.

One of the first things we did when we started working on the Space Elevator was to scrap the science fiction scenarios. These scenarios were fun and visionary but really impractical for the first Elevator. We cut everything back until we had what was required and only that. Everything sent into space is used, nothing wasted, nothing extra.

But, in building the first ribbon, we are dependent on what can be launched to GEO via rockets. No moving asteroids, the minimal ribbon as we have discussed, and current technology throughout. We also must remember that the reason we are developing the Elevator is because rockets are expensive. That affects our deployment as well. We use rockets when required but once even the smallest Elevator is operational we use it for everything. How this shakes out is that we use rockets to deploy an initial ribbon down from geosynchronous orbit and then use climbers to add more ribbon strands, increasing its capacity.

The basic scenario is that four 20-ton payloads, two ribbon, a spacecraft and fuel, are launched by rocket to LEO (low Earth orbit, where the ISS resides) and assembled. The engines are fired and the entire system moves from an altitude of 300 kilometers to geosynchronous orbit at 36,000 kilometers. Once there the ribbon deploys back down toward Earth as the engines continue to fire and move the satellite upward. Eventually the end of the ribbon splashes down in the ocean and the satellite, now at the end of its life, is hanging at the far end of the cable.

This is a stable Space Elevator that can be climbed by mini climbers. Climbers, like spiders, are attached and sent up the ribbon; each annexes a small ribbon to the first making it stronger and wider and can return to Elevator Space Port to be prepared for the next journey.

After 280 climbers over two and a half years the ribbon is large enough to support 20-ton climbers. The Elevator now consists of a ribbon a meter or more wide, extending 100,000 km with one used satellite at the upper end acting as the counterweight. The 35,500 km of ribbon between GEO and Earth is pulled by Earth gravity. The 65,000 km beyond GEO is pushed away, deflected, by centrifugal forces. Do it right and they just balance, creating the illusion of a ribbon just hanging by itself.

That phrase "a ribbon just hanging by itself" deserves expanding on.

The 100,000 km ribbon will be made up of hundreds of strands, woven together, but at first we will be laying out a ribbon made of one strand only. That single strand will be 100,000 km long but only 20 cm wide. It will be wound onto a large storage drum in the same manner that steel cables or fiber-optic cables are stored on drums today. So far so good!

Although the ribbon material is hi-tech, the winding and storing methodologies are traditional and widely used today and even that length is within the capacities of modern winding machines, which routinely handle thousands or millions of kilometers of fibre-optic cable.

The first ribbon will be wound onto a large drum and sent to GEO where it will be unrolled with the end racing back toward Earth. The end approaching Earth will have a propulsion unit; electronics and navigation beacon so that we can retrieve the end when it reaches Earth.

The central "drum" containing the ribbon will itself be encased in a housing that also combines the functions of satellite and propulsion unit. It will also contain a motor capable of inducing the drum to spin (and counteracting the inertia needed to do that). Eventually it will get incorporated into the larger Geo Station.

To get it to an orbital height of 35,500 km, we will need to use traditional rocketry to deploy the cable mission. Based on our calculations, the entire mission could be lifted by four heavy rockets launched in sequence.

It is ironic, of course, that we need a heavy-lift rocket to get the Elevator cable into space, since once the ribbon is deployed, the entire heavy-lift rocket industry will become redundant. We could be paranoid here and worry about the risk of sabotage, but let's instead be optimistic and assume that everyone understands the benefits. Perhaps an offer of a free trip to the Moon for all of the rocket employees would be an incentive?

From LEO, rockets propel the mission unit on to GEO.

Once the unit is in place at an orbit of 35,500 km, (35,786 km on average) it needs to be stabilized and oriented. Then it is ready to start paying out the cable.

18. USING THE FIRST SPACE ELEVATOR

With the ribbon in place how will it be used? We already have an existing launch market that focuses on placing telecommunication, military, and research satellites in orbit. The Elevator will be a big hit with these customers. But because of how the Elevator operates these markets will quickly expand to include new types of commercial satellites, energy systems and many new exploration spacecraft. We will address these in more detail later but be sure that there will be no shortage of customers.

Going beyond these first markets, the Elevator will enable an entirely new age in space. For this next stage we will begin placing infrastructure in orbit and on our neighboring planets. Initial infrastructure will include communications, fuel depots, food stores, repair depots, and basic space stations.

Construction will likely start on the first major space station since the ISS. The ISS itself will likely have been decommissioned by then, so there will no longer be a space station in orbit around Earth. It will not be at the end of the ribbon: for now, the only reason we have a cable 100,000 km long is that it needs to be that length in order for angular momentum to keep it rotating at the same speed as the Earth.

The first space station will be built at a much more useful place: GEO, some 35,500 km out from Earth. Here it provides a jumping-off point to launch and service all of those satellites that already sit at this height.

Since this point is the fulcrum, or gravitationally balanced center, of the Elevator system, we can build as big a space station as we like at this point, and we intend that it will indeed be very big.

The first space station, the Geo Station, will bear more resemblance to a large hotel than a broom cupboard. In addition to servicing operational satellites, by the 2030s we can offer tourists a genuine ride into space and a holiday at a space station 35,500 km from Earth, at a cost that will be affordable to many.

Meanwhile, at the far end of the cable 100,000 km away, development will be continuing. In the 2030s, a further space station, the Outer Space Station, will be in operation at this point. This will have docking facilities for rockets, as, from here, we will start a regular shuttle service to the Moon.

For reasons explained later we can't simply extend the ribbon to the Moon and attach it there. We still have to use rockets to get around in space. However, the cost of a rocket journey from the end of the ribbon, to the Moon, will be a fraction of existing rocket costs. Most of the cost of today's rocket trips lies in actually getting off of Earth. Once you are past GEO, costs are a tiny fraction of current space travel costs.

Compared to using rockets, servicing a Moon base via the Space Elevator has two advantages: travel costs are 98% lower, and the tonnage that can be shipped is higher.

With daily services to the Moon, it becomes practical to build a large Moon base, with heavy equipment that can be shipped out from Earth, fuel dumps, tractors, diggers, cranes, etc. The early stages of a Moon base will be for pioneers and construction specialists. However, soon we will be able to ship back products mined from the Moon, and tourists will be able to visit the Moon.

19. BUILDING A SECOND SPACE ELEVATOR

If a Space Elevator were so useful, and so cheap compared with rockets, you would be right in thinking that a broken one would be disastrous.

With the first Elevator in place, we finally have that cheap and easy way into space that we have been waiting for, and we don't want to lose it. Having only one Space Elevator is risky, for accidents can happen, and if the ribbon ever breaks, we have to go back to rockets to return.

The obvious answer is to build more than one ribbon, and indeed this is what we will do. We will quickly want to build a second ribbon, and the cost of a second one is so much lower, since we can use the first ribbon to take the material up into space. In fact, we envisage that Earth will eventually have many Space Elevators in operation.

So profitable

Since the cost saving is so dramatic, it should be apparent that the organization or nation that controls the first space elevator will be best positioned to profit from space business and tourism. In fact, the profits are likely to be so high as to propel this organization or nation into the position of being the dominant economic power of the 21st and 22nd Centuries.

For this reason, we can expect a surge in construction activity, as many nations seek to establish their own space elevator and start to compete in space. That is, unless the first nation to build a space elevator cable, is minded to shoot down attempts by other nations, in order to monopolize space. Sadly, the risk of space wars is there already, with President Trump wanting to establish a Space Force, and space elevators could accelerate levels of conflict. For now, we hope not.

20. TACTICAL DEPLOYMENT ISSUES

In this section we explain the deployment in simple terms addressing the typical questions that get asked.

Naming the parts

What do we call the various parts of a Space Elevator assembly? There is a series of new parts to name and shipping analogies have been utilized. They are:

Elevator Space Port (ESP) being the Earth Station, floating in the ocean on Earth.
Geo Station (GS) being the space station situated at geostationary orbit (GEO).
Outer Space Station (OSS) being the end-of-cable space station, 100,000 km out into space.

Start by thinking of the last time you went skiing in the snow. You may ride a large cable car, which travels along suspended cables. At each end, technology engages and disengages the cable car from the cable. The technology is somewhat similar, and the word "lift" was sometimes used, with variations such as space-lift, liftport and the like.

The "lift" or "elevator" analogies were the first to be used, and though various alternatives have been proposed, the space "elevator" has emerged as the main term. Think of an elevator in a building. These already travel up 50 or more levels and we enter those suspended boxes without a second thought. So, imagined the elevator continuing on for thousands of levels.

The difference, apart from the length of the journey, is the effect of gravity. For a traveler, apparent gravity shifts as they ascend. The experience starts off just like an elevator, but as apparent gravity fades and zero gravity beckons, the experience becomes one of traveling along a road or railway, only with hardly any gravity. The Space Elevator can be thought of as a track into space, with a vehicle traveling along the track between the Earth and the Outer Space Station. Imagine an airplane, where you are pinned to your seat, with gravity behind you as you take off, only to change orientation, so you sit comfortably in your seat, and the airplane is moving forward instead of up. That analogy gives the general idea.

So, what feels like a rocket capsule at the start, turns into a car or airplane further up. The cable car analogy gets used to describe the vehicle riding up the cable. Here we will simply call it the "cable car".

But again, there is a difference to a rocket capsule, which has constraints of weight and size. The car also needs to be light, not heavy, but as it is not traveling fast, especially in the atmosphere, it can be big, as big as we can engineer it. For passengers, the experience will be akin to traveling in a large railway carriage, with plenty of room for movement, but with little or no gravity. No more being stuck in the last row of seats in Economy or Coach! Plenty of room.

The ribbon stretches from Earth into space. This will be something like a sheet of paper. Imagine a roll of paper like used in newspaper production, rolled out into space, a meter or several meters wide. This functions as a track, or road. The cable car will have wheels gripping the ribbon on both faces, squeezing to get traction. We've not built a road in this configuration before, though a conveyor belt comes close.

What to call the stations or terminals at each end of the cable and elsewhere? The phrase "space station" is simple enough today: there is only one space station up there presently, being the ISS.

There are three main stations or terminals on a space elevator: At sea level on Earth, mid-way (sort of) at GEO or geostationary earth orbit, and at the far end of the ribbon, the last station.

Elevator Space Port (ESP) being the Earth Station or Port at the Earth-end of the cable, akin to JFK, Heathrow, or Cape Canaveral.

Geo Station (GS) being the space station 35,500 km along the cable, situated at geostationary orbit (GEO) and the largest space station on the cable. This will be the largest space station, eventually a Star Trek sort of thing, kilometers across.

Outer Space Station (OSS) being the end-of-cable space station, or, in whimsical moments, the FarOut Terminus, 100,000 km out into space.

Deploying the Ribbon

In the original science fiction novels, the Elevator systems were huge - too large to carry up from Earth. In these scenarios asteroids or moons were captured and moved into position and then mined for material to make the massive cables. In reality we now have a very different system, one where the ribbon and counterweight are much less massive and don't require a moon or asteroid for raw supplies. However, deployment is still a challenge.

In the present day, construction of a suspension bridge is completed by first sending a small string across a chasm and then pulling successively larger strings, and then ropes are pulled across until a full bridge is built. Similarly, we will be sending an initial "string" into space and then pulling up material to build the Elevator we need.

The total mass of the system is going to be around 1,400 tons - a large truck on the highway carries 24 tons, a rocket can carry 20 - 45 tons to LEO.

So, we still can't grab the entire system and launch it into space. But what we can do and plan to do is to launch several spools of ribbon and a spacecraft to LEO. The spacecraft is pretty standard equipment and is only a small fraction of the size of the current International Space Station.

The spools each hold 100,000 km of ribbon - not small spools, more like the size of a medium-sized truck. The spacecraft and spools are connected together and move up to GEO. At GEO the spacecraft begins to unspool the ribbon back down toward Earth. As it is unspooled the spacecraft moves upward, as a counterbalance to the effect of gravity on the ribbon. Unspooling at high speed, it will still take a couple of weeks to reach Earth where it will be secured to an anchor.

Sounds easy? Well, parts can get a little complex and we do have engineering to do but it is all well within our capabilities. Some of the complexities come in at various stages:

Deploying the ribbon "downward" when in space "down" often doesn't exist as a direction.

Catching the ribbon as it descends from space sounds tricky as well, but we only need to think back to the Apollo days where the catching was done by the ocean. We will only need to retrieve the floating end of the ribbon. Similar "catches" have recently been made when comet material was returned to the Earth.

Controlling the deployment rate is challenging since if we slow the ribbon at the top it takes seven hours for the action to ripple down the ribbon before it slows at the bottom. It will be like playing with a super rubber band.

Once we get this first ribbon connected we aren't done, only just started. This first ribbon can hold one ton and will be stable for a while but the value is that it gives us a foothold, a small piece of infrastructure that allows us to build a larger "bridge" between Earth and space.

To build a larger Elevator we will begin running our spider climbers up the first cable. Each climber will act like a spider, laying down a track of ribbon alongside the first one, strengthening it. Climber after climber will ascend the cable adding more strands of ribbon to the first one. Hundreds of climber journeys later a massive ribbon many times the size of the first, and able to carry twenty ton climbers, will be complete.

Climbers and Cars

So, that was how the ribbon stays in place. What do you do with a ribbon hanging in space? You climb it, of course. Climb it hundreds and thousands of kilometers into space to deliver payloads to orbit and do everything space enthusiasts have hoped for since Apollo.

Mechanical climbers carry out the ascent. They clamp onto the ribbon traveling much like a vertical railroad (though, as previously noted, when gravity falls away it will resemble more of a horizontal railroad as the Earth gets left behind).

The climbers would carry satellites, cargo and eventually people to space. They are straightforward mechanical machines not unlike many vehicles we have in our everyday lives.

The climbers, or cable cars as we call them, whether a goods car, repair vehicle or passenger car, consist of a central structure, a power system, a drive system, a simple control system, communications hardware, cargo holds, fuel and air tanks, passenger holds, and that is most of it. The subsystems can be largely separated and defined.

The frame or structure of a car has to support the other components in the appropriate locations for the minimal mass possible. Weight is a constraint - limited to what the ribbon can support, but volume isn't a constraint. The cable car can be very large, as long as it isn't too heavy.

The drive system is a set of rollers that clamp onto the ribbon and pull the cable car up from Earth - a challenge because no one has built such a car before. However, think of a snowmobile, a conveyor belt, a printing press with massive rollers and paper flying past, think of treadmills and the moving walkways in airports, belt sanders, and it becomes evident that we can learn from related engineering projects.

It is amazing to look at the engineering that has gone into each of these systems and how closely they resemble what we need to do in our drive system. We have adapted these technologies to answer our engineering questions. To drive the treads or rollers we will have a set of electric motors not that different from what is in electric cars though they will be modified for operation in space. Again the technology is there for us to use.

The control system and communications

In spite of the amazing things the car will do, it has a pretty simple operation - go up, stop, go up, go up faster, go down, go down slower, stop, much like a train on rails. Satellites today are many times more complex, personal computers run much more complex operations continuously, remote control cars, self driving cars and airplanes have more complex systems than we need and the communications systems aren't too much different either. This technology is readily available in off the shelf systems.

For optimal performance cable cars will be sent up at regular times and spaced to limit the total strain on the ribbon. Initially the space elevator is a one-way system at any time. You can either run cars up or cars down the ribbon at any time but not both. But with the addition of passing and overtaking technology to the cars, we will be able to run cars in both directions. Once usage reaches the point where cargo is being run down to Earth, as well as up, then this assists in managing the weight distribution logistics.

Allowing for the safe weight limits of the ribbon system, we initially estimate there will be some six cars on the ribbon at any one time. Their speed of travel will be at least 200 km per hour average (slower near the Earth, faster further away), taking almost seven days to reach the Geo Station. Beyond that, speeds can increase, allowing the Outer Space Station to be reached in another five days, making a total trip time of about twelve days. The ride down will take about the same length of time.

This may come as a surprise. We are used to seeing rockets move at speeds of 7 km/sec to 12 km/sec into space. Fast, for sure, but expensive. By comparison, the elevator cars are slow, but cheap. It takes a full day, near enough, to fly half-way around the world, say from Sydney to London, so it is understandable a trip to GEO could take a week.

The cable and climbers are the key components of the Space Elevator but there are many other essential subsystems that play important roles.

These include the Elevator Space Port, the deployment satellite, the tracking systems, the power beaming stations and all the general business activities associated with its operation.

21. RIBBON DESIGN

A number of alternative ribbon designs have been considered. The final choice will depend on proof-of-concept testing in space, once CNT ribbons have been mass produced.

R1: the very first ribbon strand

The prime choice has always been for one single interwoven ribbon in space (OSIRIS). The OSIRIS concept deploys a single ribbon all the way from Earth, through the Geo Station at GEO, to the Outer Space Port at the far end of the ribbon.

The very first ribbon (R1) sent into space will be thin and lightweight, and exists simply to put a ribbon into place, so that subsequent ribbons can be lifted along this first ribbon, instead of having to be launched by rocket. Mass becomes weight only at the Earth end. Depending on how strong CNT has become by launch date, we envisage R1 to be 20 cm wide.

Once R1 has been snagged and tethered to the Earth Space Port, further unwinding happens between GEO and the eventual end position of the ribbon, where the Outer Space Station will be located. Initially, the rocket/satellite/ribbon Combo Mass will have started unwinding downwards from GEO, but the Combo Mass will simultaneously be propelled away from GEO in the first instance, to balance the pull of gravity on R1 as it descends. With the Earth end snagged, the Combo Mass can then send the balance of the ribbon in the other direction, out beyond GEO into space, with a small counterweight (to be correct, "countermass") at the end which will, later on, become the nucleus of the OSS.

So, initially, the Combo Mass, which will later be the nucleus of the Geo Station, will actually be situated beyond GEO in order to balance the gravitational forces on the system.

Next, a small, lightweight, crawler will climb up and along R1, towing a second ribbon, R2, behind it. On R1, weight, near the Earth, is a critical factor. The crawler will need power, but there will be a trade-off between engines (of whatever power source) and speed. The power systems cannot be heavier, in weight, than R1 can cater for, allowing also the weight of R2 being towed up. So speed of travel will be limited to the capacity of the power system. But, for this first crawler, speed is not so important as ensuring it makes it to the Combo Mass safely, where R2 gets secured to R1, ready for the entire lengths to be sewed together. It might take a month for the crawler to achieve this objective.

Although the crawler will be a stripped-down version of the cable cars to come, it still requires a suite of basic systems: wheels or a locomotion system to grip and climb on R1, some form of power and engines, clamps to hold and tow R2, mechanicals to attach R2 to R1 at the Combo Mass, space hardened electronics, batteries, computing and communications systems.

In addition, the crawler will have a machine dubbed the ribbon sewing machine (RSM). Once R2 is attached to R1 at the Combo Mass, the job of the crawler will be to descend again, stitching or sewing the fringe of the second ribbon to the fringe of the first ribbon, widening it. Again, a slow job, perhaps taking several months, but by the time the crawler arrives back at the Earth Space Port, the result will be a ribbon twice as wide, with twice the lifting capacity.

We continue this process. Having doubled the weightlifting capacity of the ribbon, the next crawler can have an upgraded and heavier propulsion system, so a third ribbon can be added in less time. And so on. How many such ribbons will we lift and stitch together? It depends on the CNT strength we are working with, but based on present projections, a ribbon of between one and ten meters width would be feasible.

It is a linear trade-off. The wider the ribbon, the greater the carrying capacity. A thin ribbon has limited capacity, resulting in slow travel and limited cargo capacity per cable car. Double the ribbon, and the cable cars can have upgraded power and engine systems that are heavier in weight near the Earth yet can travel faster and/or have greater cargo capacity.

System basics were itemized above, but of course there are more systems to be incorporated. By the 2030's an AI (artificial intelligence) may be built into the computer systems. Others include risk management systems, cameras (we want to see what is going on), on-the-fly repair systems (a space AAA membership would be helpful!), radiation shielding, cargo holds, plus many of the systems that are already routine in rockets.

In addition to the boot-strap activity of adding more ribbon strands, until the ribbon has nominal capacity, the crawlers will start delivering items and mass to the Combo Mass, for assembly of the GS and the OSS. As mass gets deployed at the GS and the OSS, and remains there, the computer systems will gradually move the GS closer to its final GEO location, to maintain nominal gravitational balance.

To move the GS closer to GEO, we could wind in some of the ribbon, at the Earth Terminal, but, having gone to so much trouble to get the ribbon out of the earth gravity well, it seems a waste to just pull it back again. So one of the early deliverables will be a power/engine/locomotion assembly enabling the Combo Mass to travel along the ribbon towards GEO. As mass moves around the ribbon system, the Combo Mass, which will eventually be the GS, will always need capability to move along the ribbon, to or from Earth, to balance the system.

Notice we described it above as moving "along" the ribbon, rather than up or down. It is time to dispose of our Earth-centric view of the ribbon as being up or down. Sure, for the first third of the journey to the HS, the journey feels like "up". But, in your mind, imagine the ribbon bending until, instead of a ladder going upwards, it is a road going sideways. It will feel like a car climbing a hill, at first up a steep incline, but the road leveling out until it is flat at the top of the hill (at GEO), then becoming a gentle down-hill run between GS and the OSS. For much of the ribbon and the journey, it will approximate a road journey on a level road, so it makes sense to rotate the ribbon by ninety degrees in our mind and think of the journey into space as a road trip, sideways from Earth.

Some deliverables will form a nascent Geo Station, adding hard-ware, software and mechanical handlers to the Combo Mass, expanding it into a mini GS. In addition to mechanical handling capacity (think the equivalent of cranes unloading cargo cars, manipulators placing items into position, AI robotic machines connecting and installing items), we will start to deliver and as-semble human-rated, radiation shielded, living space, becoming the first human space station outside of the ISS in LEO (always as-suming the toted Lunar Gateway or similar hasn't been built).

At first, the ribbon between the GS and the OSS will still be a single ribbon, but at some point we commence stitching more ribbons to it, just as was done between Earth and the Combo Mass, until we have a full width ribbon stretching out to the FFAT. The logistics are different. By this point, sections of GS to the OSS rib-bon can be taken to the GS on a drum. The crawler between GS and the OSS does not have the low earth gravity well to contend with, so can be designed differently. It will drag each ribbon out from GS to the OSS (imagine this as a downhill journey, as it feels like negative gravity out here due to the centrifugal force acting on the ribbon), then stitch each to the main ribbon until the en-tire ribbon is constructed.

Now, more mass can be added to the OSS. One reason for mass is to balance the mass on the rest of the ribbon. Mass between Earth and GEO tends to fall towards Earth. Mass beyond GEO, between the GS and OSS, tends to fly away from Earth. We desire the GS to be actually at GEO and the whole system to be in gravitational balance, so we require considerable mass at the OSS to act as a balance.

The OSS mass will consist of useful things. We construct another space station at OSS, since this is going to be a jumping off point, for spaceships headed to the Moon or elsewhere, as well as a potential capture point for incoming spaceships. It will be akin to an industrial site on Earth. It will have a human rated living space, warehouses for stock, spares, cable cars and cargo, as well as fuel stores and gas stores (air, oxygen, primarily). There will be docks for arrivals and departures of spaceships. This is going to be a very busy place, a true space port, in space, connecting to other space ports at the Moon, Mars and others.

Deploying the entire space elevator system is more than about just getting a ribbon into space, although that in itself is a world-changing innovation. Getting everything built, erecting the initial Geo Station and the initial OSS putting all the systems in place, is projected to take about two years. With advances in computing and AI, this entire deployment can be done without putting any humans into space. It would be sensible to build the entire thing, before any humans contemplate riding the space elevator. This way, risk is lowered, it all gets built and tested.

However, there will be pressure to allow humans to use the space elevator, even in the early stages. Partly its the excitement, the experience, of riding the ribbon into space. Partly, and even allowing for AI and better robotics, humans may be needed in-situ, to deal with systems issues.

To provide for a human presence calls for an additional set of logistics. A cable car with human cabins, is needed, with facilities, air, water, food, space suits. Ideally, a basic human living space at the Geo Station needs to be in place. Radiation shielding is a critical factor once within and beyond the Van Allen Belts. So, apart from a pilot journey, as a proof-of-concept and, well, fun, the hardware and basic systems should be put in place prior to any human journeys.

A linear ribbon, or a loop?

As an aside, all the above assumes a single ribbon, with the cable car having the means of propulsion.

Amongst the several space elevator concepts, one is for the elevator ribbon to be deployed as a loop, which is then rotated. Imagine a conveyor belt, only 100,000 km in length. At the Earth end, powerful engines rotate, so any particular piece of the "conveyor belt" or ribbon travels up from Earth on one side, round a wheel at the far end, and back to Earth in a continuous loop.

The advantage of such a configuration is a car doesn't need its' own propulsion unit; it clamps onto the moving cable and gets pulled along. Also, the power to keep the cable moving can be generated in engines on Earth, where normal engines and fuel can be used.

The disadvantage lies in managing a rapidly moving cable in space, over 100,000 km, with the two respective cables slapping against each other, causing and suffering from friction. But if the concept can be demonstrated to work, it is a neat way of disposing of power problems.

22. THE SINGLE RIBBON CONCEPT

The core assumption is for the OSIRIS (One Single Independent Ribbon Into Space) concept, a single, wide, ribbon, into space, forming a road to the GS and the OSS.

The ribbon is stretched (and the effects of gravity really do result in "stretched") between a fixed point on Earth, the Earth Space Port, and a variable point at the OSS. As the distribution of mass changes on the ribbon, and due to the influence of gravity from, primarily, the Moon and Sun, the ribbon will oscillate, in slow motion, like a very long guitar string. So, the ribbon needs position management in order to maintain stability.

The advantage of the single ribbon system, OSIRIS, is that a single ribbon is fixed in position, for a first approximation of "fixed", and so is relatively manageable from a systems point of view.

The main disadvantage is, cars traveling on the ribbon need wheels, or similar, for locomotion, and the ability to power themselves, whether that power comes from gas, aviation fuel, nuclear, solar panels, etc. There is a lot of design work to do, just to get one cable car traveling along the ribbon, and back.

The ribbon is designed like a train line (sorry for mixing metaphors), to handle more than one car traveling on it, at a time. The journey time from Earth to the GS at GEO is expected to be around one week, and another week to the OSS, so, if we only have capacity for one cable car, we can only "launch" a car every couple of weeks, at best. This is not inconsistent with rocket launches, which tend to happen at about one or two rocket launches a week, on average, but we can do better.

The ribbon will be capable of handling many cars at a time, spaced out. Our initial assumption is for a daily launch of cars from Earth. At a car a day, the ribbon to GS will have around seven cars traveling to GS, and another seven returning from GS to Earth, making a ribbon designed to have capacity for 14 cars.

Initially, and for years to come, the cars going to GS will be loaded, the cars returning to Earth mostly empty. No problem. We are moving mass along the ribbon, which commences in possession of the angular momentum at datum (sea level) but which has to absorb additional angular momentum as it travels along the ribbon. This tends to drag the ribbon backwards in its orbit, so the ribbon will tend to drift westward at the far end. But its tethered to Earth at the Earth Space Port so, like a pendulum, after a while it will swing back towards a nominal equilibrium. The process of adding angular momentum to mass is achieved at the cost of an infinitesimal slowing of the rotation of the Earth.

Don't worry, it would take maybe a million years before the effect would be measurable. Long before then, we will be mining stuff elsewhere in space and bringing it back to Earth, along with return journeys of spaceships and so on. When we load cable cars and bring mass back to Earth, the opposite effect happens to angular momentum, so eventually we will restore the lost angular momentum.

But to return to those 14 cable cars. At some point they are going to have to pass each other. A car heading out to the Geo Station must pass a car returning from there. With one ribbon, its a bit like cars trying to pass each other on a single track farm road. We need a system, a methodology for doing this and a number have been developed.

Bear in mind, the cars can not, can never, let go of the ribbon. They must always be clamped to it, else they risk floating off into space and falling to Earth.

Recall the ribbon will become five to ten meters wide. That is as wide as a normal road, wide enough for each car to "keep right" and drive past each other. If a car is driving along one side of the ribbon, this does create issues which are dealt with below.

The alternative is for outbound bound cars to drive along one side of the ribbon, returning cars driving along the other side. Unfortunately, they still have to do a dance to get past each other. Unlike a road on Earth, where your car wheels sit on the surface, a car driving along the ribbon needs wheels or similar on both sides of the ribbon, clamped together, in order to provide the grip necessary for locomotion. So, passing maneuvers would involve the two cars connecting in some way, then moving sideways on the ribbon to get past each other, a bit messy, but feasible.

Next, we imagine outbound and inbound cars traveling along the same side of the ribbon. So, when two cars meet, a method is needed for one car to ride over the top of another. The cars could be designed (see Book Three of this series) with a rigid line across the edge of them, and the wheels can be designed so the axle system can expand beyond the width of a cable car, so that, for the wheels of the inbound car expand over the outbound car and they pass.

Cable car design is expanded on in Book Three of this series, but you can see that car design is related to ribbon design and management.

23. THE DUAL RIBBON CONCEPT

"Two ribbons in the sky" sounds like the name of a good song! "Ribbon in the Sky" is already taken, a Stevie Wonder hit in 1982, covered by Boys II Men amongst others.

So, if cars passing or overtaking is such a big deal, why not two ribbons? One ribbon will be for outbound traffic, the other for inbound traffic.

Soon after the first ribbon is constructed, an early order of business will certainly be to build a second ribbon, just in case the first ribbon becomes unusable, breaks or is destroyed by space debris or something. This is the insurance policy scenario. Having created a cheaper way to travel into space, we never want to be in the position of losing it, of having to return to rockets to build a second one.

The cost of building the second space elevator system will be far less than the first, since we don't have to use rockets. The components get sent up the first elevator system, shifted around in space, and constructed as the second system. As explained later, a second system will be built, and we identify locations sufficient to build perhaps eight space elevator systems, located around the Earth, in the Pacific, Atlantic and Indian oceans.

But the problem we are trying to solve here is that of cars passing or overtaking. So we envisage a cable car outbound on ribbon one, returning on ribbon two. The implication of this is, the cable car must be able to transfer from one cable to another, at or near the GS at GEO.

The first concept involves building two ribbons, right next to each other. Or another way of looking at it, instead of one ribbon, ten meters wide, we build two ribbons, each five meters wide, stretching from GEO to Earth.

The key problem becomes the interaction of the two ribbons through vibration. Think of two guitar strings. Each, when plucked, vibrates. Put them close enough together and they can hit each other each time they vibrate. So, two ribbons in space will both oscillate, and being next to each other, could hit each other, slap each other or worse, become entangled in some way. This will need computer modeling and testing, but this two-ribbon scenario comes with an increased risk of ribbon damage.

24. PAPER, ROCK, SCISSORS

So why not separate the two ribbons to a safe distance? We don't yet know what that distance is, but computer modeling will deliver an answer. A 35,500 km ribbon from GS to Earth may well oscillate a distance, mid-way, of tens or hundreds of kilometers.

Scenario three is the scissors scenario. Each ribbon is anchored at Earth, some distance apart. Let's say 1,000 km for the sake of this thought experiment, easy enough to do in the Pacific. But they converge and meet at GS at GEO, before parting again past GEO towards their respective OSS, forming a scissor-like shape.

This is a better scenario. The cable cars can transfer between ribbons at the Geo Station, while the ribbons are separated enough, mid-way to Earth, to avoid interacting with each other.

There are two issues to address, though.

First, those two OSS. They are a further 65,000 km out from the Geo Station. Being in a fixed position at one end only (GEO), they will oscillate to a greater extent, due to traffic on the ribbons, and the releasing and capturing of spaceships. And, not being fixed to anything, unlike the Earth end, they won't conveniently spread out in a scissor-like shape. Their orbit profile runs from their notionally fixed point at GEO, so instead, they will both travel out in an apparent straight line from the GS. In other words, between SS and beyond, the two ribbons will be next to each other, bringing us back to the problem of the two-ribbon solution above.

Second, they both meet at GS. This is great in terms of cargo management, but the weakness lies in having just one GS. If it was ever destroyed, perhaps an asteroid hitting it, then we lose both ribbons. It's a remote possibility but it has to be planned for.

25. THE SEPARATE RIBBONS CONCEPT

The fourth scenario is for ribbons which are completely separate. If they each have their own Geo Station, not physically connected, the advantage lies beyond the GS.

Each ribbon will project in a straight line from each GS, and their respective GS's will be even further apart, 100,000 km out. This offers a better solution, though it then means the cable cars have to have propulsion to be moved from one ribbon to the other, in open space.

26. GEO STATION JUNCTIONS

Within the Geo Station, there can be a ribbon junction. Thinking in terms of trains on Earth, we can build the equivalent of a train station with several lines, shunting yards, points and lines to enable movement between them. Technically it is feasible, indeed desirable, to have a system enabling cars to move from one ribbon to another.

What if the ribbons are further apart at GEO? If they don't converge on the same point at one GS, there are two scenarios. The GS will, over time, be constructed outwards (in width as viewed from Earth) along the line of GEO, with a framework of gantries and enclosed spaces (see Book Three for more detail). With no gravity to contend with, this construction could stretch for kilometers east and west. If a second ribbon was only a few kilometers away, and the gantries stretch to it, a second Geo Station can be built, and a track joining the two Geo Stations can transport cable cars from one to the other. However, this doesn't solve anything for a scissor system - the ribbons will still stay relatively close together between the GS and the end of their ribbons.

The second option, if the second ribbon is further away, is to take advantage of the lack of gravity, and safe orbit, at GEO. A cable car, with propulsion, could be moved, like a free moving satellite, from one ribbon to the next, at GEO. Technically possible though less secure.

Alternatively, if the respective GS are separate, but within a reasonable distance, say a few thousand kilometers, we can envisage a further ribbon connecting the two GS, loose, not in tension, but able to act as a guide for cable cars being sent from one to the other, and minimizing the risk of losing a cable car in space.

So the separate ribbon scenario offers a good alternative solution, obviating the need for cable cars to pass each other on the same ribbon. Or does it? Even with this solution, our cable cars will still need a mechanical solution allowing cars to pass each other, just in case. Imagine a scenario in which a cable car breaks down while traveling. We have to plan for this. It's bound to happen at times. The car behind it, and every subsequent car, will need to be able to overtake the broken car. Plus, our space AAA repair vehicle will need to travel to the broken car, dock with it, and have facilities to effect repair, or to tow the broken vehicle to GS or to Earth.

Since we still need the facility for cars to pass or overtake, does a two-ribbon scenario help? Yes it does. We will build two ribbons anyway, and for ease of traffic management, initiating a one-way system may make sense.

We say "may" because there is still another factor in traffic management, the torque introduced by the passage of each car.

Think of it this way. A car, under load, traveling outbound from Earth, is pulling on the ribbon as it climbs. This introduces a stretch moment of inertia, as the ribbon transfers a moment of inertia into the cable car. On the other hand, a car traveling inbound to Earth hardly needs engines, or even a tight grip on the cable. Gravity will pull it down, so at first it doesn't induce a stretching moment on the ribbon. It just needs to ensure proximity to the ribbon, that it doesn't separate from the ribbon and drift off into space. As it gets closer to Earth, however, braking needs to be induced to keep the car traveling at a nominal speed.

How fast can a car go, as it returns to Earth? Technically, it could free-fall until it nears Earth, traveling much faster than its outbound partners. This could make the return journey time a fraction of the outbound journey.

There are two issues here. One is, if this inbound car has to pass or overtake other cars, that has to be done at a slow speed, very slow. Second, even on the one-way ribbons described above, with a clear run, the car has to enter the atmosphere, and come to a stop at sea level.

It won't be entering the atmosphere at the speed of a rocket (a rocket/satellite has to travel fast to stay in orbit whereas our cable car is just hanging there, not in orbit) so we are not concerned about burning up on re-entry. We are talking a speed of maybe 1,000 kph max, versus a rocket re-entry speed of over 28,000 kph. It has to brake. One way is to apply friction to the ribbon via the wheel assembly, braking, which returns the moment of inertia back to the ribbon. Another, speculative way, is to separate from the ribbon at 1,000 kmh, give the car wings (and a parachute maybe) and let it fly back to Earth since it will be traveling at the same speed as an airplane.

The one thing we don't want is an out-of-control car hurtling back to Earth at great speed, having lost the ability to brake. So, tempting though a fast return is, the current concept is to keep the cars traveling at the same speed as outbound cars, about 200 kph, so braking can be continuously applied.

Here we have envisaged a number of possible ribbon deployment scenarios and the implications for cable car movements on the ribbons. These involve a relatively stationary ribbon being climbed by powered, moving cars.

What if we reversed the concept? Can we imagine a moving ribbon, with cars not needing separate power, being clamped to the ribbon?

27. THE CONVEYER BELT CONCEPT

In industrial applications, a conveyor belt system is often used to provide continuous belt movement in a direction. Applying this to the space elevator, the ribbon is woven as a continuous loop, just like a conveyor belt, and it rotates around two wheels or cylinders, one at each end. One cylinder would be at sea level, while the other would be at the GS. The ribbon then travels around the rotating cylinders, up into space and back again, in a continuous loop.

This is a neat concept, and at first glance, seems to solve a number of problems, though it introduces new problems. There are four advantages:

* The power to rotate cylinders, hence moving the ribbon, can be applied at sea level, at the Earth Space Port, which is easier than dealing with power generation in space. We can use everyday engines, perhaps diesel turbines, to do this. In addition to ease of use, it lowers the power generation costs.
* Cable cars journey along the ribbon, simply by clamping to the moving ribbon, just as, when skiing, your ski cable car clamps to the cable. At the other end, a system is in place to capture the cable car as it unclamps from the ribbon.
* Operationally it is similar to the two-ribbon concept. As the ribbon moves like a conveyor belt, there is an up ribbon and a down ribbon. If necessary, the ribbon rotation can even be reversed,

perhaps to return a cable car to Earth if there is some problem.
* Once on every rotation, every part of the ribbon returns to the
Earth Space Port. This would be useful in monitoring the ribbon
condition, and if any part of the ribbon needs repairing, the potential to do the repair on Earth, or at the Geo Station.

The cable car design may not be so different. It still needs wheels
so that the car can be eased onto the ribbon and off again, rather
than imposing a sudden load onto the ribbon. For safety purposes,
it may still need an engine for propulsion in case the ribbon has
to freeze movement for any reason. However, propulsion is most
likely needed further out from Earth, so we may not need to provide the cable car with heavier fueled engines. Perhaps the solar
panels on the car, and battery powered locomotion, would be
sufficient.

The disadvantage is a key systems management issue. This ribbon
will be a 35,500 km length conveyor belt, in tension, consisting
of 71,000 km of ribbon. How far apart will the ribbon sections
be? That is governed by the diameter of the cylinders at each
end, providing the rotation. If the diameter is measured in meters
only, then we still have the two-ribbon problem of the ribbon sections slapping or hitting each other.

In addition, because they are moving relative to each other, they
could rub against other. What would be the implications for
rubbing? Will this generate static electricity? Will this act like
sandpaper, wearing the ribbon and damaging it? The rotating cylinders will be exerting pressure on the ribbon too. Perhaps a mo-
bius strip solution would reduce any wear and tear.

Like the two-ribbon solution, computer modeling is needed to
determine how far apart the two ribbon sections need to be, in
order to obviate the risk of physical contact. If the distance is
manageable from a construction point of view, say, up to a couple
of kilometers, then the ribbon could pass around four cylinders
instead of two. Two cylinders at the Earth Space Port could be

positioned a couple of kilometers apart, same at the Geo Station end. A gantry frame can be erected at each end, to ensure the physical separation of the cylinders.

If the required separation is beyond something feasible, perhaps tens or hundreds of kilometers, it may rule out this approach.

This assumes a ribbon constructed as a conveyor belt between sea level and the GS. If this is feasible, then the section between the GS and OSS can also be a conveyor belt system. How will it be powered? The power can be generated at the GS. Alternatively, the two cylinders at GS, one being the end of the Earth-bound ribbon, one being the start of the OSS ribbon, can be connected by chains so that the entire system, both conveyor belts, are powered from Earth.

28. A SHORT CONVEYOR BELT VARIATION

A variation on the conveyor belt approach is to have a short conveyor belt at the Earth end. This model would be a conveyor belt from Earth into space, from where it attaches to a single ribbon for the rest of the journey.

The effect of gravity drops off rapidly. At 2,650 km altitude, gravity is half that of sea level. At 6,400 km altitude, gravity is a quarter that of sea level. At GEO, 35,500 km from Earth, gravity is just 0.23 m/s2

The key question is, at what altitude can we do away with fuel powered engines, and power a cable car for the rest of the journey by solar power? We don't know the answer yet, especially as advances in solar panel technology will probably change the game by the 2030s.

Suppose the answer was at 2,650 km up, for the sake of this thought experiment.

At that altitude, apparent gravity in free fall is half the force we feel at sea level, which, approximately, reduces the power load for a cable car by half. Maybe that's enough for solar power to take over.

We can imagine a junction at 2,650 km above datum, where the conveyor belt ends, which is connected in a junction to the main ribbon, with a cable car connection in place.

This is an interesting idea, in a way. By reducing the conveyor belt to a length of 2,650 km we reduce the problems of slapping and friction, wear and tear. It also provides a cheaper way of launching cable cars from Earth, to an altitude where the apparent weight has halved.

29. OTHER LESS LIKELY IDEAS

A number of ribbon alternatives have appeared over the past decade, mostly in science fiction and in our view, that's where they'll stay, since they offer no advantages and plenty of disadvantages.

Story lines in fiction have included the concept of a giant wheel, 35,500 km in diameter, rotating, where cars jump on at Earth and jump off at GEO. The problems? Where to start. There is no feasible material that would hold its shape at this size. It would skim a few thousand km of the surface of the Earth and the various forces acting on it would twist it out of shape and cause its collapse.

Another idea is for a short tether to the Earth, suspended from a satellite in space. Technically, a satellite could hold up a ribbon or tether, but then the real problems start. A satellite in LEO would be orbiting the Earth at around 26,500 kph, so how does anyone jump aboard? Or, if the satellite is geostationary, it needs rockets burning continuously to keep it up in space. Which means fuel has to be taken up to it, which means a rocket would be more effective anyway. Many of the science fiction writers adopting this idea haven't done the math and assume such a satellite would just stay in place. (This is a plea for science fiction writers to do the math, check the facts.)

Another idea is the rotating giant wheel, again, but this one has a diameter of only a few thousand kilometers, skimming Earth.

Non-starter - it can't possibly stay up in space, the whole thing will be unstable and fall to Earth.

The next variation in sci-fi is, ok, we can't have a wheel, what about a couple of slingshots rotating a central axle. Each slingshot ends in a platform, timed to drop down to Earth to seemingly pause at some point, where the characters hop on board and get whisked into space. Again, there's nothing to keep this system in space. Plus, trying to manage a slingshot, thousands of kilometers long, and position it to drop neatly, exactly, to the surface of the Earth, and by that we mean, exactly on it, not a hundred meters too low, not a hundred meters too high ... even with computer controlled AI systems, it just isn't going to happen.

All of these can make for good, or even bad, science fiction, but we are focused on technology and systems that can be modeled and demonstrated to actually work - and work in another decade, not in some far-flung future!

What makes the preferred single ribbon system practical?

* The ribbon is anchored to the Earth at a predictable and fixed location.
* This means we can use centrifugal force, out beyond GEO, to stretch a ribbon and keep in place.
* It creates useable space at GEO, with apparent zero gravity, which means we can build a GEO base, the Geo Station, as big as we like. This can be Star Trek or Star Wars size, kilometers across, as big as we can build it.
* It creates a far-out terminus, the Outer Space Station, which, benefiting from centrifugal force, is primed to launch spaceships into space, to the Moon, Mars or anywhere else. It's no good having a concept that just gets something off of Earth - you need a systems methodology to send it somewhere useful.

30. THE RIBBON MANAGEMENT SYSTEM

It isn't enough to just place a ribbon into space. Systems are needed to manage it. The ribbon management system will be computer controlled from the Earth Space Port.

We want to know where the ribbon is - exactly. Plus, cable cars need a system to track the center of the ribbon for travel purposes.

For this, we envisage a hardware bar, every kilometer or so, along the ribbon. Clamped to the ribbon, flat so cable cars can travel over them, they will have three transponders. In the center of the ribbon, a transponder provides a narrowcast of its location, used by cable cars for centering position as they travel along the ribbon.

On each side of the ribbon will be a set of sensors. One will broadcast its location so, back on Earth, the 3-D coordinates of every kilometer of ribbon can be tracked. Also will be video cameras, pointing both ways, enabling visual checks of the ribbon and the traffic on it.

Location tracking is vital. As traffic moves along the ribbon, it affects the mass distribution and causes local oscillations. Any impacts from space debris will do the same. The ribbon management system needs this information to make position adjustments to keep the ribbon in balance and to keep the center of gravity at GEO.

An asteroid or meteor hit will show up on the management system immediately, betrayed by a sudden sideways movement of the ribbon at the impact point. With this warning, that ribbon section can be inspected and repaired promptly.This also acts as an early warning system. If the ribbon experiences a shear and breaks, the sensors provide immediate information on this, enabling emergency and recovery actions.

A ribbon of this length has a natural oscillation period of about 7.2 hours. (Think of it as a very long guitar string being plucked to vibrate.) Each time the Earth rotates and the ribbon swings past the Moon it is pulled on slightly by the lunar gravity. This will cause oscillations, but it happens in slow motion, and each Earth day the impact of the Moon happens out of phase.

These sensors need power, envisaged to come from their own set of small solar panels. They need to transmit data. Although we can't run a power cable along the ribbon, it is feasible to run optical fiber for data communications, which could be used by the sensors. But the more likely scenario is for each sensor to form a wifi network of their own, using IoT (Internet of Things) technology, so they use each other to pass information along, to Earth or to the GS, where the GS uses more powerful transmitters to stay in communication with Earth. The cable cars themselves will also be in wifi contact with the system, using more powerful transmitters.

The video cameras provide a visual means of observing the ribbon, picking up any visual signs of damage, providing visual confirmation of traffic data.

In addition to cable cars, both for passengers and for goods, we will have repair and analysis vehicles, whose job is to crawl along the ribbon, looking for breakages large and small, providing high definition video and undergoing onsite repairs. This job will be automated, however, maybe there will be times when a human repair crew is necessary. The surrounding video cameras and sensors will enable real-time monitoring of activity.

In addition to these repeating systems, we can position specific hardware at points along the ribbon. For example, we could position a space radar system to provide additional means of tracking space debris and satellites.

Security from attack is vital. As the most valuable piece of infrastructure on and off the planet, it shares the problems of military bases, airports and any other vital infrastructure, so complete security systems will be in place.

31. A COMPLETED SPACE ELEVATOR RIBBON

In this book, we've demonstrated the various concepts and systems involved in constructing a space elevator ribbon. It would be great if the technical problems are solved and the first ribbon is put into space in another decade, in 2029, as first envisaged in our 2006 book Leaving the Planet by Space Elevator, or in the 2030s.

However, all timetables slip, especially when it comes to space. Assuming the CNT production problem is solved sometime during the 2020s, we may well see a decade of Earth-based applications and testing after that, such as building suspension bridges out of CNT on Earth.

A space elevator would follow on from that, but the follow-up will be rapid. Once we have proof-of-concept, and experience of CNT usage on Earth, the value of the first space elevator will be obvious to all and we envisage construction of it to commence, perhaps in the early to mid 2030s.

So, with the first ribbon deployed, how do we operate the space elevator as a system? What will the impact be, on space travel, operations, Moon bases? These themes are followed up in Book Three of this series.

THE SPACE ELEVATOR 2020 SERIES

Book Two: Construction of the Space Elevator

Published 2020

Linda J. Phillips

The Space Elevator 2020 series
Book Two:
Construction of the Space Elevator

Publisher contact: info@21stcentury.space
FaceBook www.facebook.com/lindyjaniceAuthor
Twitter @_lindaphillips
Amazon author page https://www.amazon.com/author/linda-janicephillips
Web 21stcentury.space
Web links utilized in this publication were correct at the time of writing, but they can change over time.

Published by Linda Phillips

Edwards & Phillips